Dig, Dig
Leslie Wood

Oxford University Press

Oxford Toronto Melbourne

Dig, dig.

Dig, dig.

Dig, dig.

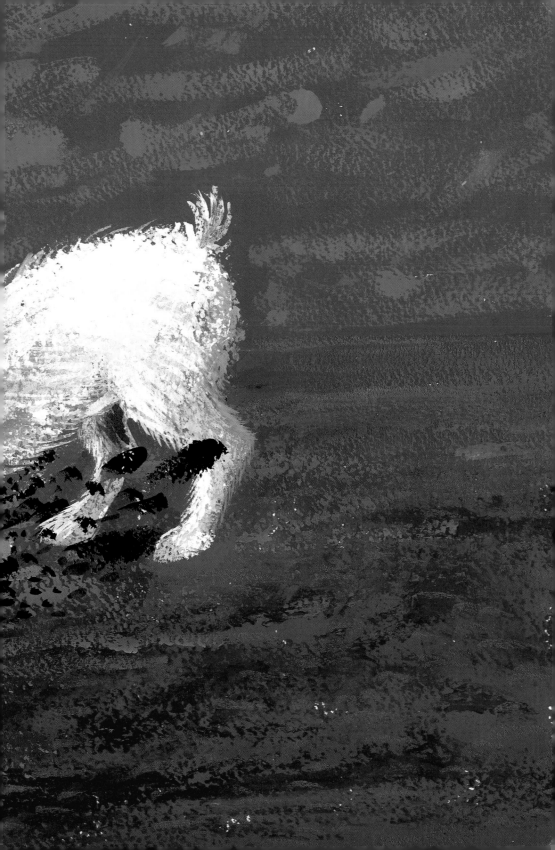

Dig, dig.

Dig, dig.

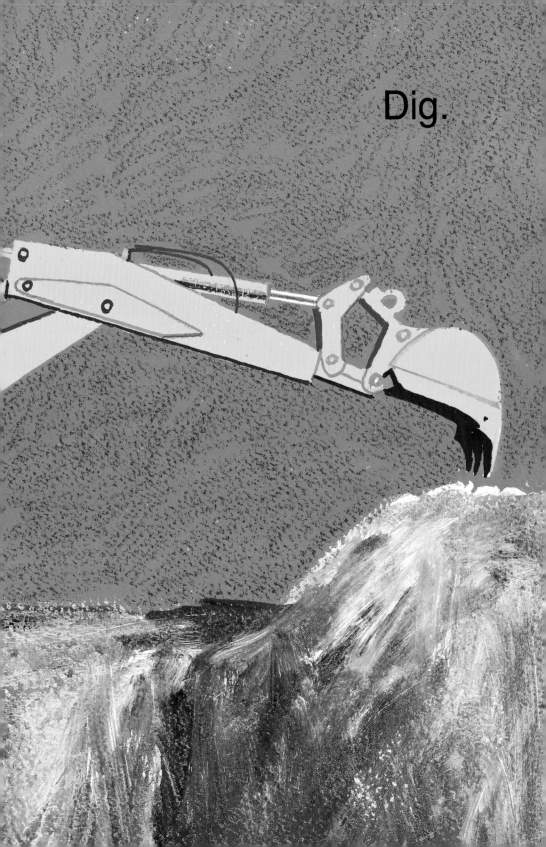

Dig.

Dig.

Oxford University Press, Great Clarendon Street, Oxford OX2 6DP

Oxford University Press – Education
198 Madison Avenue, New York, New York 10016

Oxford New York
Athens Auckland Bangkok Bogotá Buenos Aires
Cape Town Chennai Dar es Salaam Delhi
Florence Hong Kong Istanbul Karachi Kolkata Kuala Lumpur
Madrid Melbourne Mexico City Mumbai Nairobi Paris
São Paulo Shanghai Singapore Taipei Tokyo Toronto Warsaw

and associated companies in
Berlin Ibadan

Oxford is a trade mark of Oxford University Press

© Leslie Wood 1988
First published 1988
Reprinted 1991, 1994, 1995, 1997, 1998, 1999, 2001

This edition is also available in
Oxford Reading Tree Branch Library Stage 1 Pack **A**
ISBN 0 19 272100 3

British Library Cataloguing in Publication Data

Wood, Leslie
Dig, dig.—(Cat on the mat series)
I. Title   II. Series
823′.914[J]   PZ7

ISBN 0 19 272185 2
USA ISBN 0 19 849019 4

Typeset by PGY Graphic Design, Oxford
Printed in China